Tunnel Engi

A Museum Treatment

Robert M. Vogel

Alpha Editions

This edition published in 2024

ISBN : 9789362514998

Design and Setting By
Alpha Editions
www.alphaedis.com
Email - info@alphaedis.com

Figure 1.—MINING BY EARLY EUROPEAN CIVILIZATIONS, using fire setting and hand chiseling to break out ore and rock. MHT model—¾" scale. (Smithsonian photo 49260-H.)

Robert M. Vogel

TUNNEL ENGINEERING—A MUSEUM TREATMENT

During the years from 1830 to 1900, extensive developments took place in the field of tunneling, which today is an important, firmly established branch of civil engineering. This paper offers a picture of its growth from the historical standpoint, based on a series of models constructed for the Hall of Civil Engineering in the new Museum of History and Technology. The eight models described highlight the fundamental advances which have occurred between primitive man's first systematic use of fire for excavating rock in mining, and the use in combination of compressed air, an iron lining, and a movable shield in a subaqueous tunnel at the end of the 19th century.

THE AUTHOR: Robert M. Vogel is curator of heavy machinery and civil engineering, in the Smithsonian Institution's Museum of History and Technology.

Introduction

With few exceptions, civil engineering is a field in which the ultimate goal is the assemblage of materials into a useful structural form according to a scientifically derived plan which is based on various natural and man-imposed conditions. This is true whether the result be, for example, a dam, a building, a bridge, or even the fixed plant of a railroad. However, one principal branch of the field is based upon an entirely different concept. In the engineering of tunnels the utility of the "structure" is derived not from the bringing together of elements but from the separation of one portion of naturally existing material from another to permit passage through a former barrier.

In tunneling hard, firm rock, this is practically the entire compass of the work: breaking away the rock from the mother mass, and, coincidently, removing it from the workings. The opposite extreme in conditions is met in the soft-ground tunnel, driven through material incapable of supporting itself above the tunnel opening. Here, the excavation of the tunneled substance is of relatively small concern, eclipsed by the problem of preventing the surrounding material from collapsing into the bore.

Figure 2.—HOOSAC TUNNEL. METHOD OF WORKING EARLY SECTIONS of the project; blast holes drilled by hand jacking. MHT model—½" scale. (Smithsonian photo 49260-L.)

In one other principal respect does tunnel engineering differ widely from its collateral branches of civil engineering. Few other physical undertakings are approached with anything like the uncertainty attending a tunnel work. This is even more true in mountain tunnels, for which test borings frequently cannot be made to determine the nature of the material and the geologic conditions which will be encountered.

The course of tunnel work is not subject to an overall preliminary survey; the engineer is faced with not only the inability to anticipate general contingencies common to all engineering work, but with the peculiar and often overwhelming unpredictability of the very basis of his work.

Subaqueous and soft-ground work on the other hand, while still subject to many indeterminates, is now far more predictable than during its early history, simply because the nature of the adverse condition prevailing eventually was understood to be quite predictable. The steady pressures of earth and water to refill the excavated area are today overcome with relative ease and consistency by the tunneler.

In tunneling as in no other branch of civil engineering did empiricism so long resist the advance of scientific theory; in no other did the "practical engineer" remain to such an extent the key figure in establishing the success or failure of a project. The Hoosac Tunnel, after 25 years of legislative, financial, and technical difficulties, in 1875 was finally driven to successful completion only by the efforts of a group who, while in the majority were trained civil engineers, were to an even greater extent men of vast practical ability, more at home in field than office.

DeWitt C. Haskin (see p. 234), during the inquest that followed the death of a number of men in a blowout of his pneumatically driven Hudson River Tunnel in 1880, stated in his own defense: "I am not a scientific engineer, but a practical one ... I know nothing of mathematics; in my experience I have grasped such matters as a whole; I believe that the study of mathematics in that kind of work [tunneling] has a tendency to dwarf the mind rather than enlighten it...." An extreme attitude perhaps, and one which by no means adds to Haskin's stature, but a not unusual one in tunnel work at the time. It would not of course be fair to imply that such men as Herman Haupt, Brunel the elder, and Greathead were not accomplished theoretical engineers. But it was their innate ability to evaluate and control the overlying physical conditions of the site and work that made possible their significant contributions to the development of tunnel engineering.

Tunneling remained largely independent of the realm of mathematical analysis long after the time when all but the most insignificant engineering works were designed by that means. Thus, as structural engineering has advanced as the result of a flow of new theoretical concepts, new, improved, and strengthened materials, and new methods of fastening, the progress of tunnel engineering has been due more to the continual refinement of constructional techniques.

A NEW HALL OF CIVIL ENGINEERING

In the Museum of History and Technology has recently been established a Hall of Civil Engineering in which the engineering of tunnels is comprehensively treated from the historical standpoint—something not previously done in an American museum. The guiding precept of the exhibit has not been to outline exhaustively the entire history of tunneling, but rather to show the fundamental advances which have occurred between primitive man's first systematic use of fire for excavating rock in mining, and the use in combination of compressed air, iron lining, and a movable shield in a subaqueous tunnel at the end of the 19th century. This termination date was selected because it was during the period from about 1830 to 1900 that the most concentrated development took place, and during which tunneling became a firmly established and important branch of civil engineering and indeed, of modern civilization. The techniques of present-day tunneling are so fully related in current writing that it was deemed far more useful to devote the exhibit entirely to a segment of the field's history which is less commonly treated.

Figure 3.—HOOSAC TUNNEL. WORKING OF LATER STAGES with Burleigh pneumatic drills mounted on carriages. The bottom heading is being drilled in preparation for blasting out with nitroglycerine. MHT model—½" scale. (Smithsonian photo 49260-M.)

The major advances, which have already been spoken of as being ones of technique rather than theory, devolve quite naturally into two basic classifications: the one of supporting a mass of loose, unstable, pressure-exerting material—soft-ground tunneling; and the diametrically opposite problem of separating rock from the basic mass when it is so firm and solid that it can support its own overbearing weight as an opening is forced through it—rock, or hard-ground tunneling.

To exhibit the sequence in a thorough manner, inviting and capable of easy and correct interpretation by the nonprofessional viewer, models offered the only logical means of presentation. Six tunnels were selected, all driven in the 19th century. Each represents either a fundamental, new concept of tunneling technique, or an important, early application of one. Models of these works form the basis of the exhibit. No effort was made to restrict the work to projects on American soil. This would, in fact, have been quite impossible if an accurate picture of tunnel technology was to be drawn; for as in virtually all other areas of technology, the overall development in this field has been international. The art of mining was first developed highly in the Middle Ages in the Germanic states; the tunnel shield was invented by a Frenchman residing in England, and the use of compressed air to exclude the water from subaqueous tunnels was first introduced on a major work by an American. In addition, the two main subdivisions, rock and soft-ground tunneling, are each introduced by a model not of an actual working, but of one typifying early classical methods which were in use for centuries until the comparatively recent development of more efficient systems of earth support and rock breaking. Particular attention is given to accuracy of detail throughout the series of eight models; original sources of descriptive and graphic information were used in their construction wherever possible. In all cases except the introductory model in the rock-tunneling series, representing copper mining by early civilizations, these sources were contemporary accounts.

The plan to use a uniform scale of reduction throughout, in order to facilitate the viewers' interpretation, unfortunately proved impractical, due to the great difference in the amount of area to be encompassed in different models, and the necessity that the cases holding them be of uniform height. The related models of the Broadway and Tower Subways represent short sections of tunnels only 8 feet or so in diameter enabling a relatively large scale, 1½ inches to the foot, to be used. Conversely, in order that the model of Brunel's Thames Tunnel be most effective, it was necessary to include one of the vertical terminal shafts used in its construction. These were about 60 feet in depth, and thus the much smaller scale of ¼ inch to the foot was used. This variation is not as confusing as might be thought, for the human figures in each model provide an immediate and positive sense of proportion and scale.

Careful thought was devoted to the internal lighting of the models, as this was one of the critical factors in establishing, so far as is possible in a model, an atmosphere convincingly representative of work conducted solely by artificial light. Remarkable realism was achieved by use of plastic rods to conduct light to the tiny sources of tunnel illumination, such as the candles on the miners' hats in the Hoosac Tunnel, and the gas lights in the Thames Tunnel. No overscaled miniature bulbs, generally applied in such cases, were used. At several points where the general lighting within the tunnel proper has been kept at a low level to simulate the natural atmosphere of the work, hidden lamps can be operated by push-button in order to bring out detail which otherwise would be unseen.

The remainder of the material in the Museum's tunneling section further extends the two major aspects of tunneling. Space limitations did not permit treatment of the many interesting ancillary matters vital to tunnel engineering, such as the unique problems of subterranean surveying, and the extreme accuracy required in the triangulation and subsequent guidance of the boring in long mountain tunnels; nor the difficult problems of ventilating long workings, both during driving and in service; nor the several major methods developed through the years for driving or constructing tunnels in other than the conventional manner. [1]

Rock Tunneling

While the art of tunneling soft ground is of relatively recent origin, that of rock tunneling is deeply rooted in antiquity. However, the line of its development is not absolutely direct, but is more logically followed through a closely related branch of technology—mining. The development of mining techniques is a practically unbroken one, whereas there appears little continuity or relationship between the few works undertaken before about the 18th century for passage through the earth.

The Egyptians were the first people in recorded history to have driven openings, often of considerable magnitude, through solid rock. As is true of all major works of that nation, the capability of such grand proportion was due solely to the inexhaustible supply of human power and the casual evaluation of life. The tombs and temples won from the rock masses of the Nile Valley are monuments of perseverance rather than technical skill. Neither the Egyptians nor any other peoples before the Middle Ages have left any consistent evidence that they were able to pierce ground that would not support itself above the opening as would firm rock. In Egypt were established the methods of rock breaking that were to remain classical until the first use of gun-powder blasting in the 17th century which formed the basis of the ensuing technology of mining.

Notwithstanding the religious motives which inspired the earliest rock excavations, more constant and universal throughout history has been the incentive to obtain the useful and decorative minerals hidden beneath the earth's surface. It was the miner who developed the methods introduced by the early civilizations to break rock away from the primary mass, and who added the refinements of subterranean surveying and ventilating, all of which were later to be assimilated into the new art of driving tunnels of large diameter. The connection is the more evident from the fact that tunnelmen are still known as miners.

COPPER MINING, B.C.

Therefore, the first model of the sequence, reflecting elemental rock-breaking techniques, depicts a hard-rock copper mine (fig. 1). Due to the absence of specific information about such works during the pre-Christian eras, this model is based on no particular period or locale, but represents in a general way, a mine in the Rio Tinto area of Spain where copper has been extracted since at least 1000 B.C. Similar workings existed in the Tirol as early as about 1600 B.C. Two means of breaking away the rock are shown: to the left is the most primitive of all methods, the hammer and chisel, which require no further description. At the right side, the two figures are shown utilizing the first rock-breaking method in which a force beyond that of human muscles was employed, the age-old “fire-setting” method. The rock was thoroughly heated by a fierce fire built against its face and then suddenly cooled by dashing water against it. The thermal shock disintegrated the rock or ore into bits easily removable by hand.

Figure 4.—HOOSAC TUNNEL. Bottom of the central shaft showing elevator car and rock skip; pumps at far right. In the center, the top bench is being drilled by a single column-mounted Burleigh drill. MHT model—½" scale. (Smithsonian photo 49260-N.)

The practice of this method below ground, of course, produced a fearfully vitiated atmosphere. It is difficult to imagine whether the smoke, the steam, or the toxic fumes from the roasting ore was the more distressing to the miners. Even when performed by labor considered more or less expendable, the method could be employed only where there was ventilation of some sort: natural chimneys and convection currents were the chief sources of air circulation. Despite the drawbacks of the fire system, its simplicity and efficacy weighed so heavily in its favor that its history of use is unbroken almost to the present day. Fire setting was of greatest importance during the years of intensive mining in Europe before the advent of explosive blasting, but its use in many remote areas hardly slackened until the early 20th century because of its low cost when compared to powder. For this same reason, it did have limited application in actual tunnel work until about 1900.

Direct handwork with pick, chisel and hammer, and fire setting were the principal means of rock removal for centuries. Although various wedging systems were also in favor in some situations, their importance was so slight that they were not shown in the model.

HOOSAC TUNNEL

It was possible in the model series, without neglecting any major advancement in the art of rock tunneling, to complete the sequence of development with only a single additional model. Many of the greatest works of civil engineering have been those concerned directly with transport, and hence are the product of the present era, beginning in the early 19th century. The development of the ancient arts of route location, bridge construction, and tunnel driving received a powerful stimulation after 1800 under the impetus of the modern canal, highway, and, especially, the railroad.

The Hoosac Tunnel, driven through Hoosac Mountain in the very northwest corner of Massachusetts between 1851 and 1875, was the first major tunneling work in the United States. Its importance is due not so much to this as to its being literally the fountainhead of modern rock-tunneling technology. The remarkable thing is that the work was begun using methods of driving almost unchanged during centuries previous, and was completed twenty years later by techniques which were, for the day, almost totally mechanized. The basic pattern of operation set at Hoosac, using pneumatic rock drills and efficient explosives, remains practically unchanged today.

The general history of the Hoosac project is so thoroughly recorded that the briefest outline of its political aspects will suffice here. Hoosac Mountain was the chief obstacle in the path of a railroad projected between Greenfield, Massachusetts, and Troy, New York. The line was launched by a group of Boston merchants to provide a direct route to the rapidly developing West, in competition with the coastal routes via New York. The only route economically reasonable included a tunnel of nearly five miles through the mountain—a length absolutely without precedent, and an immense undertaking in view of the relatively primitive rock-working methods then available.

Figure 5.—BURLEIGH ROCK DRILL, improved model of about 1870, mounted on frame for surface work. (Catalog and price list: The Burleigh Rock Drill Company, 1876.)

The bore's great length and the desire for rapid exploitation inspired innovation from the outset of the work. The earliest attempts at mechanization, although ineffectual and without influence on tunnel engineering until many years later, are of interest. These took the form of several experimental machines of the "full area" type, intended to excavate the entire face of the work in a single operation by cutting one or more concentric grooves in the rock. The rock remaining between the grooves was to be blasted out. The first such machine tested succeeded in boring a 24-foot diameter opening for 10 feet before its total failure. Several later machines proved of equal merit. [2] It was the Baltimore and Ohio's eminent chief engineer, Benjamin H. Latrobe, who in his *Report on the Hoosac Tunnel* (Baltimore, Oct. 1, 1862, p. 125) stated that such apparatus contained in its own structure the elements of failure, " ... as they require the machines to do too much and the powder too little of the work, thus contradicting the fundamental principles upon which all labor-saving machinery is framed ... I could only look upon it as a misapplication of mechanical genius."

Figure 6.—HOOSAC TUNNEL. Flash-powder photograph of Burleigh drills at the working face. (*Photo courtesy of State Library, Commonwealth of Massachusetts.*)

Latrobe stated the basic philosophy of rock-tunnel work. No mechanical agent has ever been able to improve upon the efficiency of explosives for the shattering of rock. For this reason, the logical application of machinery to tunneling was not in replacing or altering the fundamental process itself, but in enabling it to be conducted with greater speed by mechanically drilling the blasting holes to receive the explosive.

Actual work on the Hoosac Tunnel began at both ends of the tunnel in about 1854, but without much useful effect until 1858 when a contract was let to the renowned civil engineer and railroad builder, Herman Haupt of Philadelphia. Haupt immediately resumed investigations of improved tunneling methods, both full-area machines and mechanical rock drills. At this time mechanical rock-drill technology was in a state beyond, but not far beyond, initial experimentation. There existed one workable American machine, the Fowle drill, invented in 1851. It was steam-driven, and had been used in quarry work, although apparently not to any commercial extent. However, it was far too large and cumbersome to find any possible application in tunneling. Nevertheless, it contained in its operating principle, the seed of a practical rock drill in that the drill rod was attached directly to and reciprocated by a double-acting steam piston. A point of great importance was the independence of its operation on gravity, permitting drilling in any direction.

While experimenting, Haupt drove the work onward by the classical methods, shown in the left-hand section of the model (fig. 2). At the far right an advance heading or adit is being formed by pick and hammer work; this is then deepened into a top heading with enough height to permit hammer drilling, actually the basic tunneling operation. A team is shown "double jacking," i.e., using two-handed hammers, the steel held by a third man. This was the most efficient of the several hand-drilling methods. The top-heading plan was followed so that the bulk of the rock could be removed in the form of a bottom bench, and the majority of drilling would be downward, obviously the most effective direction. Blasting was with black powder and its commercial variants. Some liberty was taken in depicting these steps so that both operations might be shown within the scope of the model: in practice the heading was kept between 400 and 600 feet in advance of the bench so that heading blasts would not interfere with the bench work. The bench carriage simply facilitated handling of the blasted rock. It was rolled back during blasts.

Figure 7.—HOOSAC TUNNEL. GROUP OF MINERS descending the west shaft with a Burleigh drill. (*Photo courtesy of State Library, Commonwealth of Massachusetts.*)

The experiments conducted by Haupt with machine drills produced no immediate useful results. A drill designed by Haupt and his associate, Stuart Gwynn, in 1858 bored hard granite at the rate of $^{5}/_{8}$ inch per minute, but was not substantial enough to bear up in service. Haupt left the work in 1861, victim of intense political pressures and totally unjust accusations of corruption and mismanagement. The work was suspended until taken over by a state commission in 1862. Despite frightful ineptitude and very real corruption, this period was exceedingly important in the long history both of Hoosac Tunnel and of rock tunneling in general.

The merely routine criticism of the project had by this time become violent due to the inordinate length of time already elapsed and the immense cost, compared to the small portion of work completed. This served to generate in the commission a strong sense of urgency to hurry the project along. Charles S. Storrow, a competent engineer, was sent to Europe to report on the progress of tunneling there, and in particular on mechanization at the Mont Cenis Tunnel then under construction between France and Italy. Germain Sommeiller, its chief engineer, had, after experimentation similar to Haupt's, invented a reasonably efficient drilling machine which had gone into service at Mont Cenis in March 1861. It was a distinct improvement over hand drilling, almost doubling the drilling rate, but was complex and highly unreliable. Two hundred drills were required to keep 16 drills at work. But the vital point in this was the fact that Sommeiller drove his drills not with steam, but air, compressed at the tunnel portals and piped to the work face. It was this single factor, one of application rather than invention, that made the mechanical drill feasible for tunneling.

All previous effort in the field of machine drilling, on both sides of the Atlantic, had been directed toward steam as the motive power. In deep tunnels, with ventilation already an inherent problem, the exhaust of a steam drill into the atmosphere was inadmissible. Further, steam could not be piped over great distances due to serious losses of energy from radiation of heat, and condensation. Steam generation within the tunnel itself was obviously out of the question. It was the combination of a practical drill, and the parallel invention by Sommeiller of a practical air compressor that resulted in the first workable application of machine rock drilling to tunneling.

Figures 8 & 9.—HOOSAC TUNNEL. CONTEMPORARY ENGRAVINGS. As such large general areas could not be sufficiently illuminated for photography, the Museum model was based primarily on artists' versions of the work. (*Science Record*, 1872; *Leslie's Weekly*, 1873.)

The Sommeiller drills greatly impressed Storrow, and his report of November 1862 strongly favored their adoption at Hoosac. It is curious however, that not a single one was brought to the U.S., even on trial. Storrow does speak of Sommeiller's intent to keep the details of the machine to himself until it had been further improved, with a view to its eventual exploitation. The fact is, that although workable, the Sommeiller drill proved to be a dead end in rock-drill development because of its many basic deficiencies. It did exert the indirect influence of inspiration which, coupled with a pressing need for haste, led to renewed trials of drilling machinery at Hoosac. Thomas Doane, chief engineer under the state commission, carried this program forth with intensity, seeking and encouraging inventors, and himself working on the problem. The pattern of the Sommeiller drill was generally followed; that is, the drill was designed as a separate, relatively light mechanical element, adapted for transportation by several miners, and attachable to a movable frame or carriage during operation. Air was of course the presumed power. To be effective, it was necessary that a drill automatically feed the drill rod as the hole deepened, and also rotate the rod automatically to maintain a round, smooth hole. Extreme durability was essential, and usually proved the source of a machine's failure. The combination of these characteristics into a machine capable of driving the drill rod into the rock with great force, perhaps five times per second, was a severe test of ingenuity and materials. Doane in 1864 had three different experimental drills in hand, as well as various steam and water-powered compressors.

Success finally came in 1865 with the invention of a drill by Charles Burleigh, a mechanical engineer at the well-known Putnam Machine Works of Fitchburg, Massachusetts. The drills were first applied in the east heading in June of 1866. Although working well, their initial success was limited by lack of reliability and a resulting high expense for repairs. They were described as having "several weakest points." In November, these drills were replaced by an improved Burleigh drill which was used with total success to the end of the work. The era of modern rock tunneling was thus launched by Sommeiller's insight in initially applying pneumatic power to a machine drill, by Doane's persistence in searching for a thoroughly practical drill, and by Burleigh's mechanical talent in producing one. The desperate need to complete the Hoosac Tunnel may reasonably be considered the greatest single spur to the development of a successful drill.

The significance of this invention was far reaching. Burleigh's was the first practical mechanical rock drill in America and, in view of its dependability, efficiency, and simplicity when compared to the Sommeiller drill, perhaps in the world. The Burleigh drill achieved success almost immediately. It was placed in production by Putnam for the Burleigh Rock Drill Company before completion of Hoosac in 1876, and its use spread throughout the western mining regions and other tunnel works. For a major invention, its adoption was, in relative terms, instantaneous. It was the prototype of all succeeding piston-type drills, which came to be known generically as "burleighs," regardless of manufacture. Walter Shanley, the Canadian contractor who ultimately completed the Hoosac, reported in 1870, after the drills had been in service for a sufficient time that the techniques for their most efficient use were fully understood and effectively applied, that the Burleigh drills saved about half the drilling costs over hand drilling. The per-inch cost of machine drilling averaged 5.5 cents, all inclusive, vs. 11.2 cents for handwork. The more important point, that of speed, is shown by the reports of average monthly progress of the tunnel itself, before and after use of the air drills.

Year	*Average monthly progress in feet*
1865	55
1866	48
1867	99
1868	—
1869	138
1870	126
1871	145
1872	124

Figure 10.—TRINITROGLYCERINE BLAST at Hoosac Tunnel. (*Leslie's Weekly*, 1873.)

The right portion of the model (fig. 3) represents the workings during the final period. The bottom heading system was generally used after the Burleigh drills had been introduced. Four to six drills were mounted on a carriage designed by Doane. These drove the holes for the first blast in the center of the heading in about six hours. The full width of the heading, the 24-foot width of the tunnel, was then drilled and blasted out in two more stages. As in the early section, the benches to the rear were later removed to the full-tunnel height of about 20 feet. This operation is shown by a single drill (fig. 4) mounted on a screw column. Three 8-hour shifts carried the work forward: drilling occupied half the time and half was spent in running the carriage back, blasting, and mucking (clearing the broken rock).

The tunnel's 1028-foot central shaft, completed under the Shanley contract in 1870 to provide two additional work faces as well as a ventilation shaft is shown at the far right side of this half of the model. Completed so near the end of the project, only 15 percent of the tunnel was driven from the shaft.

The enormous increase in rate of progress was not due entirely to machine drilling. From the outset of his jurisdiction, Doane undertook experiments with explosives as well as drills, seeking an agent more effective than black powder. In this case, the need for speed was not the sole stimulus. As the east and west headings advanced further and further from the portals, the problem of ventilation grew more acute, and it became increasingly difficult to exhaust the toxic fumes produced by the black powder blasts.

In 1866, Doane imported from Europe a sample of trinitroglycerine, the liquid explosive newly introduced by Nobel, known in Europe as "glonoïn oil" and in the United States as "nitroglycerine." It already had acquired a fearsome reputation from its tendency to decompose with heat and age and to explode with or without the slightest provocation. Nevertheless, its tremendous power and characteristic of almost complete smokelessness led Doane to employ the chemist George W. Mowbray, who had blasted for Drake in the Pennsylvania oil fields, to develop techniques for the bulk manufacture of the new agent and for its safe employment in the tunnel.

Figure 11.—HOOSAC TUNNEL survey crew at engineering office. The highest accuracy of the aboveground and underground survey work was required to insure proper vertical and horizontal alignment and meeting of the several separately driven sections. (*Photo courtesy of State Library, Commonwealth of Massachusetts.*)

Mowbray established a works on the mountain and shortly developed a completely new blasting practice based on the explosive. Its stability was greatly increased by maintaining absolute purity in the manufacturing process. Freezing the liquid to reduce its sensitivity during transport to the headings, and extreme caution in its handling further reduced the hazard of its use. At the heading, the liquid was poured into cylindrical cartridges for placement in the holes. As with the Burleigh drill, the general adoption of nitroglycerine was immediate once its qualities had been demonstrated. The effect on the work was notable. Its explosive characteristics permitted fewer blast holes over a given frontal area of working face, and at the same time it was capable of effectively blowing from a deeper drill hole, 42 inches against 30 inches for black powder, so that under ideal conditions 40 percent more tunnel length was advanced per cycle of operations. A new fuse and a system of electric ignition were developed which permitted simultaneous detonation and resulted in a degree of effectiveness impossible with the powder train and cord fusing used with the black powder. Over a million pounds of nitroglycerine were produced by Mowbray between 1866 and completion of the tunnel.

Figure 12.—WORKS AT THE CENTRAL SHAFT, HOOSAC TUNNEL, for hoisting, pumping and air compressing machinery, and general repair, 1871. *(Photo courtesy of State Library, Commonwealth of Massachusetts.)*

Figure 13.—HOOSAC TUNNEL. AIR-COMPRESSOR BUILDING on Hoosac River near North Adams. The compressors were driven partially by waterpower, derived from the river. *(Photo courtesy of State Library, Commonwealth of Massachusetts.)*

Figure 14.—West portal of Hoosac Tunnel before completion, 1868, showing six rings of lining brick. (*Photo courtesy of State Library, Commonwealth of Massachusetts.*)

When the Shanleys took the work over in 1868, following political difficulties attending operation by the State, the period of experimentation was over. The tunnel was being advanced by totally modern methods, and to the present day the overall concepts have remained fundamentally unaltered: the Burleigh piston drill has been replaced by the lighter hammer drill; the Doane drill carriage by the more flexible "jumbo";nitroglycerine by its more stable descendant dynamite and its alternatives; and static-electric blasting machines by more dependable magnetoelectric. But these are all in the nature of improvements, not innovations.

Unlike the preceding model, there was good documentation for this one. Also, the Hoosac was apparently the first American tunnel to be well recorded photographically. Early flashlight views exist of the drills working at the heading (fig. 6) as well as of the portals, the winding and pumping works at the central shaft, and much of the machinery and associated aspects of the project. These and copies of drawings of much of Doane's experimental apparatus, a rare technological record, are preserved at the Massachusetts State Library.

Soft-Ground Tunneling

So great is the difference between hard-rock and soft-ground tunneling that they constitute two almost separate branches of the field. In penetrating ground lacking the firmness or cohesion to support itself above an opening, the miner's chief concern is not that of removing the material, but of preventing its collapse into his excavation. The primitive methods depending upon brute strength and direct application of fire and human force were suitable for assault on rock, but lacked the artifice needed for delving into less stable material. Roman engineers were accomplished in spanning subterranean ways with masonry arches, but apparently most of their work was done by cut-and-cover methods rather than by actual mining.

Not until the Middle Ages did the skill of effectively working openings in soft ground develop, and not until the Renaissance was this development so consistently successful that it could be considered a science.

RENAISSANCE MINING

From the earliest periods of rock working, the quest for minerals and metals was the primary force that drove men underground. It was the technology of mining, the product of slow evolution over the centuries, that became the technology of the early tunnel, with no significant modification except in size of workings.

Every aspect of 16th century mining is definitively detailed in Georgius Agricola's remarkable *De re Metallica*, first published in Basel in 1556. During its time of active influence, which extended for two centuries, it served as the authoritative work on the subject. It remains today an unparalleled early record of an entire branch of technology. The superb woodcuts of mine workings and tools in themselves constitute a precise description of the techniques of the period, and provided an ideal source of information upon which to base the first model in the soft-ground series.

Figure 15.—CENTERING FOR PLACEMENT OF FINISHED STONEWORK at west portal, 1874. At top-right are the sheds where the lining brick was produced. *(Photo courtesy of State Library, Commonwealth of Massachusetts.)*

The model, representing a typical European mine, demonstrates the early use of timber frames or "sets" to support the soft material of the walls and roof. In areas of only moderate instability, the sets alone were sufficient to counteract the earth pressure, and were spaced according to the degree of support required. In more extreme conditions, a solid lagging of small poles or boards was set outside the frames, as shown in the model, to provide absolute support of the ground. Details of the framing, the windlass, and all tools and appliances were supplied by Agricola, with no need for interpretation or interpolation.

The basic framing pattern of sill, side posts and cap piece, all morticed together, with lagging used where needed, was translated unaltered into tunneling practice, particularly in small exploratory drifts. It remained in this application until well into the 20th century.

The pressure exerted upon tunnels of large area was countered during construction by timbering systems of greater elaboration, evolved from the basic one. By the time that tunnels of section large enough to accommodate canals and railways were being undertaken as matter-of-course civil engineering works, a series of nationally distinguishable systems had emerged, each possessing characteristic points of favor and fault. As might be suspected, the English system of tunnel timbering, for instance, was rarely applied on the Continent, nor were the German, Austrian or Belgian systems normally seen in Great Britain. All were used at one time or another in this country, until the American system was introduced in about 1855. While the timbering commonly remained in place in mines, it would be followed up by permanent masonry arching and lining in tunnel work.

Overhead in the museum Hall of Civil Engineering are frames representing the English, Austrian and American systems. Nearby, a series of small relief models (fig. 19) is used to show the sequence of enlargement in a soft-ground railroad tunnel of about 1855, using the Austrian system. Temporary timber support of tunnels fell from use gradually after the advent of shield tunneling in conjunction with cast-iron lining. This formed a perfect support immediately behind the shield, as well as the permanent lining of the tunnel.

Figure 16.—WEST PORTAL UPON COMPLETION, 1876. (*Photo courtesy of New-York Historical Society.*) Click on image for a color version of poster.

BRUNEL'S THAMES TUNNEL

The interior surfaces of tunnels through ground merely unstable are amenable to support by various systems of timbering and arching. This becomes less true as the fluidity of the ground increases. The soft material which normally comprises the beds of rivers can approach an almost liquid condition resulting in a hydraulic head from the overbearing water sufficient to prevent the driving of even the most carefully worked drift, supported by simple timbering. The basic defect of the timbering systems used in mining and tunneling was that there was inevitably a certain amount of the face or ceiling unsupported just previous to setting a frame, or placing over it the necessary section of lagging. In mine work, runny soil could, and did, break through such gaps, filling the working. For this reason, there were no serious attempts made before 1825 to drive subaqueous tunnels.

In that year, work was started on a tunnel under the Thames between the Rotherhithe and Wapping sections of London, under guidance of the already famous engineer Marc Isambard Brunel (1769-1849), father of I. K. Brunel. The undertaking is of great interest in that Brunel employed an entirely novel apparatus of his own invention to provide continuous and reliable support of the soft water-bearing clay which formed the riverbed. By means of this "shield," Brunel was able to drive the world's first subaqueous tunnel. [3]

The shield was of cast-iron, rectangular in elevation, and was propelled forward by jackscrews. Shelves at top, bottom, and sides supported the tunnel roof, floor, and walls until the permanent brick lining was placed. The working face, the critical area, was supported by a large number of small "breasting boards," held against the ground by small individual screws bearing against the shield framework. The shield itself was formed of 12 separate frames, each of which could be advanced independently of the others. The height was 22 feet 3 inches: the width 37 feet 6 inches.

The progress was piecemeal. In operation the miners would remove one breasting board at a time, excavate in front of it, and then replace it in the advanced position—about 6 inches forward. This was repeated with the next board above or below, and the sequence continued until the ground for the entire height of one of the 12 sections had been removed. The board screws for that section were shifted to bear on the adjacent frames, relieving the frame of longitudinal pressure. It could then be screwed forward by the amount of advance, the screws bearing to the rear on the completed masonry. Thus, step by step the tunnel progressed slowly, the greatest weekly advance being 14 feet.

Figure 17.—SOFT-GROUND TUNNELING. The support of walls and roof of mine shaft by simple timbering; 16th century. MHT model—¾" scale. (Smithsonian photo 49260-J.)

In the left-hand portion of the model is the shaft sunk to begin operations; here also is shown the bucket hoist for removing the spoil. The V-type steam engine powering the hoist was designed by Brunel. At the right of the main model is an enlarged detail of the shield, actually an improved version built in 1835.

The work continued despite setbacks of every sort. The financial ones need no recounting here. Technically, although the shield principle proved workable, the support afforded was not infallible. Four or five times the river broke through the thin cover of silt and flooded the workings, despite the utmost caution in excavating. When this occurred, masses of clay, sandbags, and mats were dumped over the opening in the riverbed to seal it, and the tunnel pumped out. I. K. Brunel acted as superintendent and nearly lost his life on a number of occasions. After several suspensions of work resulting from withdrawal or exhaustion of support, one lasting seven years, the work was completed in 1843.

Despite the fact that Brunel had, for the first time, demonstrated a practical method for tunneling in firm and water-bearing ground, the enormous cost of the work and the almost overwhelming problems encountered had a discouraging effect rather than otherwise. Not for another quarter of a century was a similar project undertaken.

The Thames Tunnel was used for foot and light highway traffic until about 1870 when it was incorporated into the London Underground railway system, which it continues to serve today. The roofed-over top sections of the two shafts may still be seen from the river.

A number of contemporary popular accounts of the tunnel exist, but one of the most thorough and interesting expositions on a single tunnel work of any period is Henry Law's *A Memoir of the Thames Tunnel*, published in 1845-1846 by John Weale. Law, an eminent civil engineer, covers the work in incredible detail from its inception until the major suspension in late 1828 when slightly more than half completed. The most valuable aspect of his record is a series of plates of engineering drawings of the shield and its components, which, so far as is known, exist nowhere else. These formed the basis of the enlarged section of the shield, shown to the right of the model of the tunnel itself. A vertical section through the shield is reproduced here from Law for comparison with the model (figs. 21 and 23).

Figure 18.—SOFT-GROUND TUNNELING. The model of a 16th century mine in the Museum of History and Technology was constructed from illustrations in such works as G. E. von Löhneyss' *Bericht vom Bergwerck*, 1690, as well as the better known ones from *De re Metallica.*

Figure 19.—THE SUCCESSIVE STAGES in the enlargement of a mid-19th century railroad tunnel, using the Austrian system of timbering. MHT model.

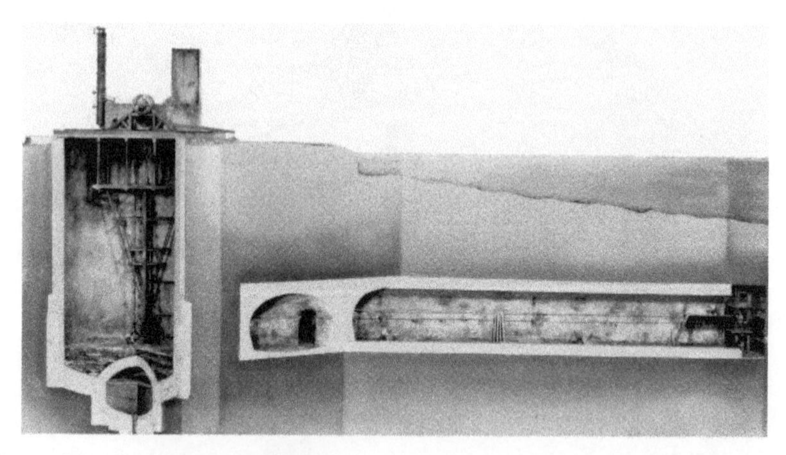

Figure 20.—M. I. BRUNEL'S THAMES TUNNEL, 1825-1843, the first driven beneath a body of water.
MHT model—¼" scale. (Smithsonian photo 49260-F.)

Figure 21.—ENLARGED DETAIL of Brunel's tunneling shield, vertical section. The first two and part of the third of the twelve frames are shown. To the left is the tunnel's completed brick lining and to the right, the individual breasting boards and screws for supporting the face. The propelling screws are seen at top and bottom, bearing against the lining. Three miners worked in each frame, one above the other. MHT model—¾" scale. (Smithsonian photo 49260-G.)

Various inventors attempted to improve upon the Brunel shield, aware of the fundamental soundness of the shield principle. Almost all bypassed the rectangular sectional construction used in the Thames Tunnel, and took as a starting point a sectional shield of circular cross section, advanced by Brunel in his original patent of 1818. James Henry Greathead (1844-1896), rightfully called the father of modern subaqueous tunneling, surmised in later years that Brunel had chosen a rectangular configuration for actual use, as one better adapted to the sectional type of shield. The English civil engineer, Peter W. Barlow, in 1864 and 1868 patented a circular shield, of one piece, which was the basis of one used by him in constructing a small subway of 1350 feet beneath the Thames in 1869, the first work to follow the lead of Brunel. Greathead, acting as Barlow's contractor, was the designer of the shield actually used in the work, but it was obviously inspired by Barlow's patents.

The reduction of the multiplicity of parts in the Brunel shield to a single rigid unit was of immense advantage and an advance perhaps equal to the shield concept of tunneling itself. The Barlow-Greathead shield was like the cap of a telescope with a sharpened circular ring on the front to assist in penetrating the ground. The diaphragm functioned, as did Brunel's breasting boards, to resist the longitudinal earth pressure of the face, and the cylindrical portion behind the diaphragm bore the radial pressure of roof and walls. Here also for the first time, a permanent lining formed of cast-iron segments was used, a second major advancement in soft-ground tunneling practice. Not only could the segments be placed and bolted together far more rapidly than masonry lining could be laid up, but unlike the green masonry, they could immediately bear the full force of the shield-propelling screws.

Barlow, capitalizing on Brunel's error in burrowing so close to the riverbed, maintained an average cover of 30 feet over the tunnel, driving through a solid stratum of firm London clay which was virtually impervious to water. As the result of this, combined with the advantages of the solid shield and the rapidly placed iron lining, the work moved forward at a pace and with a facility in startling contrast to that of the Thames Tunnel, although in fairness it must be recalled that the face area was far less.

The clay was found sufficiently sound that it could be readily excavated without the support of the diaphragm, and normally three miners worked in front of the shield, digging out the clay and passing it back through a doorway in the plate. This could be closed in case of a sudden settlement or break in. Following excavation, the shield was advanced 18 inches into the excavated area by means of 6 screws, and a ring of lining segments 18 inches in length bolted to the previous ring under cover of the overlapping rear skirt of the shield. The small annular space left between the outside of the lining and the clay by the thickness and clearance of the skirt—about an inch—was filled with thin cement grout. The tunnel was advanced 18 inches during each 8-hour shift. The work continued around the clock, and the 900-foot river section was completed in only 14 weeks. [4] The entire work was completed almost without incident in just under a year, a remarkable performance for the world's second subaqueous tunnel.

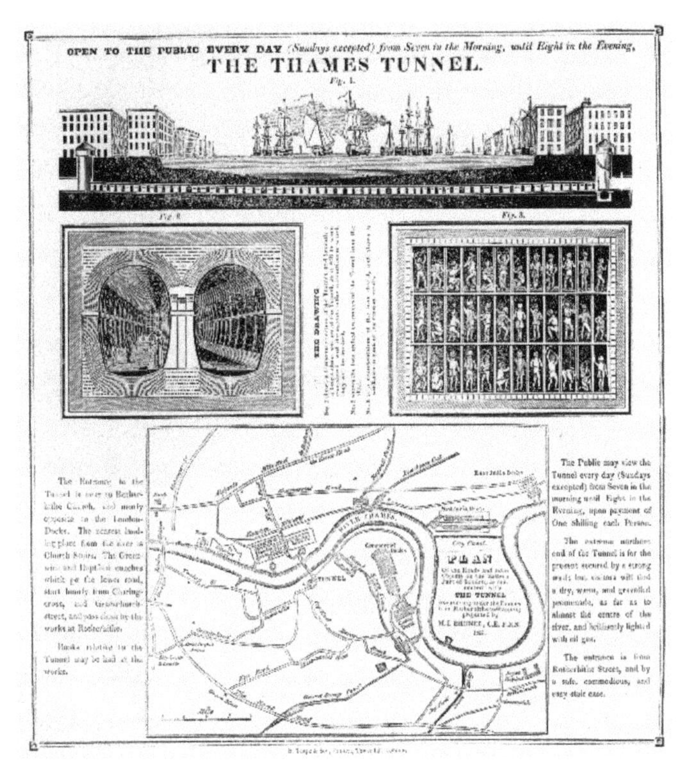

Figure 22.—BROADSIDE PUBLISHED AFTER COMMENCEMENT OF WORK on the Thames Tunnel, 1827.
(MHT collections.) Transcription of the text is presented in the Transcriber's Notes below.

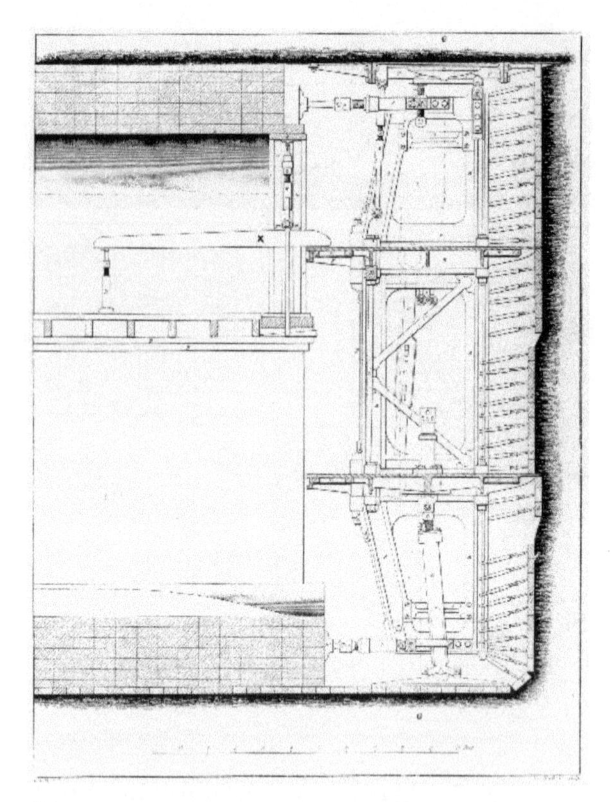

Figure 23.—VERTICAL SECTION THROUGH BRUNEL'S SHIELD. The long lever, x, supported the wood centering for turning the masonry arches of the lining. (LAW, *A Memoir of the Thames Tunnel.*)

Figure 24.—THAMES TUNNEL. SECTION THROUGH riverbed and tunnel following one of the break-throughs of the river. Inspection of the damage with a diving bell. (BEAMISH, *A Memoir of the Life of Sir Marc Isambard Brunel.*)

The Tower Subway at first operated with cylindrical cars that nearly filled the 7-foot bore; the cars were drawn by cables powered by small steam engines in the shafts. This mode of power had previously been used in passenger service only on the Greenwich Street elevated railway in New York. Later the cars were abandoned as unprofitable and the tunnel turned into a footway (fig. 32). This small tunnel, the successful driving due entirely to Greathead's skill, was the forerunner of the modern subaqueous tunnel. In it, two of the three elements essential to such work thereafter were first applied: the one-piece movable shield of circular section, and the segmental cast-iron lining.

The documentation of this work is far thinner than for the Thames Tunnel. The most accurate source of technical information is a brief historical account in Copperthwaite's classic *Tunnel Shields and the Use of Compressed Air in Subaqueous Works*, published in 1906. Copperthwaite, a successful tunnel engineer, laments the fact that he was able to turn up no drawing or original data on this first shield of Greathead's, but he presents a sketch of it prepared in the Greathead office in 1895, which is presumably a fair representation (fig. 33). The Tower Subway model was built on the basis of this and several woodcuts of the working area that appeared contemporaneously in the illustrated press. In this and the adjacent model of Beach's Broadway Subway, the tunnel axis has been placed on an angle to the viewer, projecting the bore into the case so that the complete circle of the working face is included for a more suggestive effect. This was possible because of the short length of the work included.

Henry S. Drinker, also a tunnel engineer and author of the most comprehensive work on tunneling ever published, treats rock tunneling in exhaustive detail up to 1878. His notice of what he terms "submarine tunneling" is extremely brief. He does, however, draw a most interesting comparison between the first Thames Tunnel, built by Brunel, and the second, built by Greathead 26 years later:

FIRST THAMES TUNNEL	SECOND THAMES TUNNEL (TOWER SUBWAY)
Brickwork lining, 38 feet wide by 22½ feet high.	Cast-iron lining of 8 feet outside diameter.
120-ton cast-iron shield, accommodating 36 miners.	2½-ton, wrought-iron shield, accommodating at most 3 men.
Workings filled by irruption of river five times.	"Water encountered at almost any time could have been gathered in a stable pail."
Eighteen years elapsed between start and finish of work.	Work completed in about eleven months.
Cost: $3,000,000.	Cost: $100,000.

Figure 25.—TRANSVERSE SECTION THROUGH SHIELD, after inundation. Such disasters, as well as the inconsistency of the riverbed's composition, seriously disturbed the alignment of the shield's individual sections. (LAW, *A Memoir of the Thames Tunnel.*)

Figure 26.—LONGITUDINAL SECTION THROUGH THAMES TUNNEL after sandbagging to close a break in the riverbed. The tunnel is filled with silt and water. (LAW, *A Memoir of the Thames Tunnel.*)

Figure 27.—INTERIOR OF THE THAMES TUNNEL shortly after completion in 1843. (*Photo courtesy of New York Public Library Picture Collection.*)

Figure 28.—THAMES TUNNEL in use by London Underground railway. (*Illustrated London News*, 1869?)

Figure 29.—PLACING A segment of cast-iron lining in Greathead's Tower Subway, 1869. To the rear is the shield's diaphragm or bulkhead. MHT model—1½" scale. (Smithsonian photo 49260-B.)

BEACH'S BROADWAY SUBWAY

Almost simultaneously with the construction of the Tower Subway, the first American shield tunnel was driven by Alfred Ely Beach (1826-1896). Beach, as editor of the *Scientific American* and inventor of, among other things, a successful typewriter as early as 1856, was well known and respected in technical circles. He was not a civil engineer, but had become concerned with New York's pressing traffic problem (even then) and as a solution, developed plans for a rapid-transit subway to extend the length of Broadway. He invented a shield as an adjunct to this system, solely to permit driving of the tunnel without disturbing the overlying streets.

An active patent attorney as well, Beach must certainly have known of and studied the existing patents for tunneling shields, which were, without exception, British. In certain aspects his shield resembled the one patented by Barlow in 1864, but never built. However, work on the Beach tunnel started in 1869, so close in time to that on the Tower Subway, that it is unlikely that there was any influence from that source. Beach had himself patented a shield, in June 1869, a two-piece, sectional design that bore no resemblance to the one used. His subway plan had been first introduced at the 1867 fair of the American Institute in the form of a short plywood tube through which a small, close-fitting car was blown by a fan. The car carried 12 passengers. Sensing opposition to the subway scheme from Tammany, in 1868 Beach obtained a charter to place a small tube beneath Broadway for transporting mail and small packages pneumatically, a plan he advocated independently of the passenger subway.

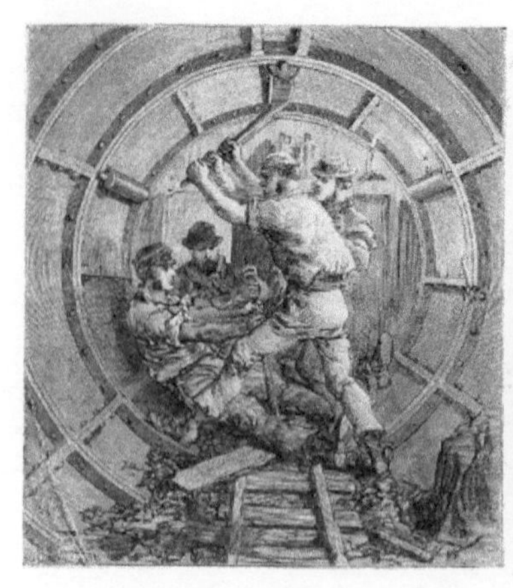

ADVANCING THE SHIELD. FITTING THE CASTINGS.

Figure 30.— CONTEMPORARY ILLUSTRATIONS of Tower Subway works used as basis of the model in the Museum of History and Technology. (*Illustrated London News*, 1869.)

Figure 31.—EXCAVATION IN FRONT OF SHIELD, Tower Subway. This was possible because of the stiffness of the clay encountered. MHT model—front of model shown in fig. 29. (Smithsonian photo 49260-A.)

Under this thin pretense of legal authorization, the sub-rosa excavation began from the basement of a clothing store on Warren Street near Broadway. The 8-foot-diameter tunnel ran eastward a short distance, made a

90-degree turn, and thence southward under Broadway to stop a block away under the south side of Murray Street. The total distance was about 312 feet. Work was carried on at night in total secrecy, the actual tunneling taking 58 nights. At the Warren Street terminal, a waiting room was excavated and a large Roots blower installed for propulsion of the single passenger car. The plan was similar to that used with the model in 1867: the cylindrical car fitted the circular tunnel with only slight circumferential clearance. The blower created a plenum within the waiting room and tunnel area behind the car of about 0.25 pounds per square inch, resulting in a thrust on the car of almost a ton, not accounting for blowby. The car was thus blown along its course, and was returned by reversing the blower's suction and discharge ducts to produce an equivalent vacuum within the tunnel.

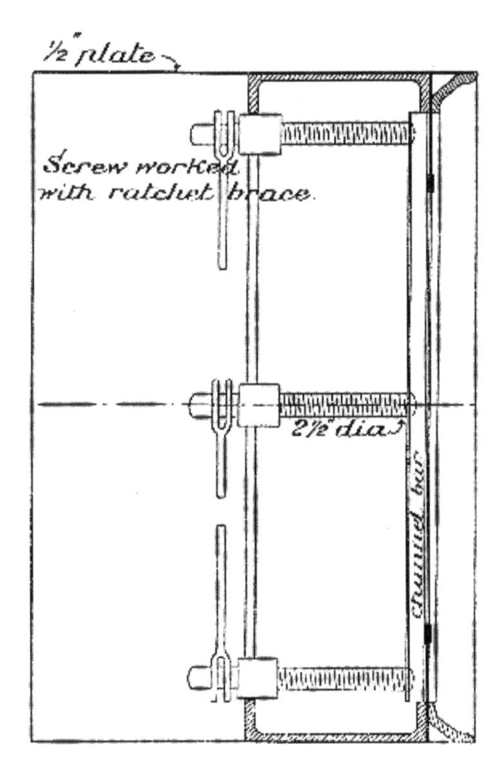

Figure 33.—VERTICAL SECTION through the Greathead shield used at the Tower Subway, 1869. The first one-piece shield of circular section. (COPPERTHWAITE, *Tunnel Shields and the Use of Compressed Air in Subaqueous Works.*)

The system opened in February of 1870 and remained in operation for about a year. Beach was ultimately subdued by the hostile influences of Boss

Tweed, and the project was completely abandoned. Within a very few more years the first commercially operated elevated line was built, but the subway did not achieve legitimate status in New York until the opening of the Interborough line in 1904. Ironically, its route traversed Broadway for almost the length of the island.

Figure 32.—INTERIOR OF COMPLETED TOWER SUBWAY. (THORNBURY, *Old and New London, 1887, vol. 1, p. 126.*)

The Beach shield operated with perfect success in this brief trial, although the loose sandy soil encountered was admittedly not a severe test of its qualities. No diaphragm was used; instead a series of 8 horizontal shelves with sharpened leading edges extended across the front opening of the shield. The outstanding feature of the machine was the substitution for the propelling screws used by Brunel and Greathead of 18 hydraulic rams, set around its circumference. These were fed by a single hand-operated pump, seen in the center of figure 34. By this means the course of the shield's forward movement could be controlled with a convenience and precision not attainable with screws. Vertical and horizontal deflection was achieved by throttling the supply of water to certain of the rams, which could be individually controlled, causing greater pressure on one portion of the shield than another. This system has not changed in the ensuing time, except, of course, in the substitution of mechanically produced hydraulic pressure for hand.

Figure 34.—BEACH'S Broadway Subway. Advancing the shield by hydraulic rams, 1869. MHT model—1½" scale. (Smithsonian photo 49260-E.)

Unlike the driving of the Tower Subway, no excavation was done in front of the shield. Rather, the shield was forced by the rams into the soil for the length of their stroke, the material which entered being supported by the shelves. This was removed from the shelves and hauled off. The ram plungers then were withdrawn and a 16-inch length of the permanent lining built up within the shelter of the shield's tail ring. Against this, the rams bore for the next advance. Masonry lining was used in the straight section; cast-iron in the curved. The juncture is shown in the model.

Figure 36.—INTERIOR of Beach Subway showing iron lining on curved section and the pneumatically powered passenger car. View from waiting room. (*Scientific American*, March 5, 1870.)

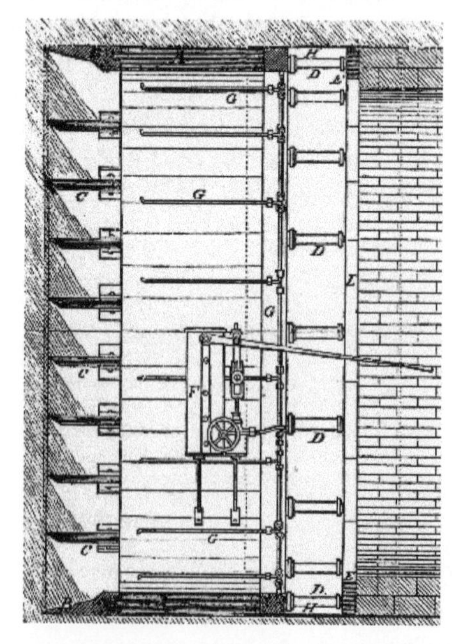

Figure 35.—VERTICAL SECTION through the Beach shield used on the Broadway Subway, showing the horizontal shelves (C), iron cutting ring

(B), hydraulic rams (D), hydraulic pump (F), and rear protective skirt (H). (*Scientific American*, March 5, 1870.)

Enlarged versions of the Beach shield were used in a few tunnels in the Midwest in the early 1870's, but from then until 1886 the shield method, for no clear reason, again entered a period of disuse finding no application on either side of the Atlantic despite its virtually unqualified proof at the hands of Greathead and Beach. Little precise information remains on this work. The Beach system of pneumatic transit is described fully in a well-illustrated booklet published by him in January 1868, in which the American Institute model is shown, and many projected systems of pneumatic propulsion as well as of subterranean and subaqueous tunneling described. Beach again (presumably) is author of the sole contemporary account of the Broadway Subway, which appeared in *Scientific American* following its opening early in 1870. Included are good views of the tunnel and car, of the shield in operation, and, most important, a vertical sectional view through the shield (fig. 35).

It is interesting to note that optical surveys for maintenance of the course apparently were not used. The article illustrated and described the driving each night of a jointed iron rod up through the tunnel roof to the street, twenty or so feet above, for "testing the position."

THE FIRST HUDSON RIVER TUNNEL

Despite the ultimate success of Brunel's Thames Tunnel in 1843, the shield in that case afforded only moderately reliable protection because of the fluidity of the soil driven through, and its tendency to enter the works through the smallest opening in the shield's defense. An English doctor who had made physiological studies of the effects on workmen of the high air pressure within diving bells is said to have recommended to Brunel in 1828 that he introduce an atmosphere of compressed air into the tunnel to exclude the water and support the work face.

This plan was first formally described by Sir Thomas Cochrane (1775-1860) in a British patent of 1830. Conscious of Brunel's problems, he proposed a system of shaft sinking, mining, and tunneling in water-bearing materials by filling the excavated area with air sufficiently above atmospheric pressure to prevent the water from entering and to support the earth. In this, and his description of air locks for passage of men and materials between the atmosphere and the pressurized area, Cochrane fully outlined the essential features of pneumatic excavation as developed since.

Figure 37.— THE GIANT ROOTS LOBE-TYPE BLOWER used for propelling the car.

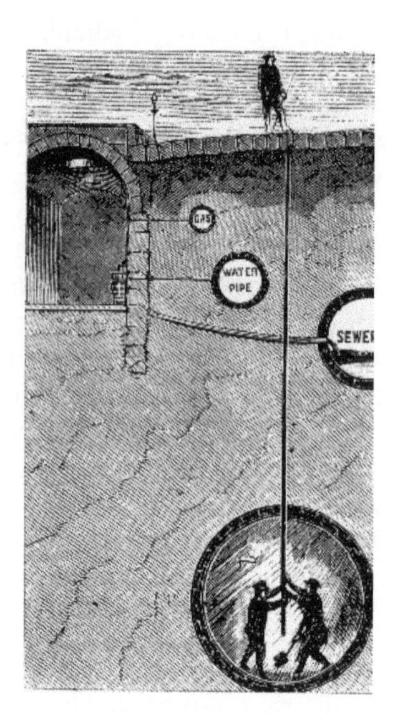

Figure 38.— TESTING ALIGNMENT of the Broadway Subway at night by driving a jointed rod up to street level. (*Scientific American*, March 5, 1870.)

In 1839, a French engineer first used the system in sinking a mine shaft through a watery stratum. From then on, the sinking of shafts, and somewhat later the construction of bridge pier foundations, by the pneumatic method became almost commonplace engineering practice in Europe and America. Not until 1879 however, was the system tried in tunneling work, and then, as with the shield ten years earlier, almost simultaneously here and abroad. The first application was in a small river tunnel in Antwerp, only 5 feet in height. This project was successfully completed relying on compressed air alone to support the earth, no shield being used. The importance of the work cannot be considered great due to its lack of scope.

In 1871 Dewitt C. Haskin (1822-1900), a west coast mine and railroad builder, became interested in the pneumatic caissons then being used to found the river piers of Eads' Mississippi River bridge at St. Louis. In apparent total ignorance of the Cochrane patent, he evolved a similar system for tunneling water-bearing media, and in 1873 proposed construction of a tunnel through the silt beneath the Hudson to provide rail connection between New Jersey and New York City.

Figure 39.—HASKIN'S pneumatically driven tunnel under the Hudson River, 1880. In the engine room at top left was the machinery for hoisting, generating electricity for lighting, and air compressing. The air lock is seen in the wall of the brick shaft. MHT model—0.3" scale. (Smithsonian photo 49260.)

Figure 40.—ARTIST'S CONCEPTION OF MINERS escaping into the air lock during the blowout in Haskin's tunnel.

It would be difficult to imagine a site more in need of such communication. All lines from the south terminated along the west shore of the river and the immense traffic—cars, freight and passengers—was carried across to Manhattan Island by ferry and barge with staggering inconvenience and at enormous cost. A bridge would have been, and still is, almost out of the question due not only to the width of the crossing, but to the flatness of both banks. To provide sufficient navigational clearance (without a drawspan), impracticably long approaches would have been necessary to obtain a permissibly gentle grade.

Haskin formed a tunneling company and began work with the sinking of a shaft in Hoboken on the New Jersey side. In a month it was halted because of an injunction by, curiously, the D L & W Railroad, who feared for their vast investment in terminal and marine facilities. Not until November of 1879 was the injunction lifted and work again commenced. The shaft was completed and an air lock located in one wall from which the tunnel proper was to be carried forward. It was Haskin's plan to use no shield, relying solely on the pressure of compressed air to maintain the work faces and prevent the entry of water. The air was admitted in late December, and the first large-scale pneumatic tunneling operation launched. A single 26-foot, double-track bore was at first undertaken, but a work face of such diameter proved unmanageable and two oval tubes 18 feet high by 16 feet wide were substituted, each to carry a single track. Work went forward with reasonable facility, considering the lack of precedent. A temporary entrance was formed of sheet-iron rings from the air lock down to the tunnel grade, at which point the permanent work of the north tube was started. Immediately behind the excavation at the face, a lining of thin wrought-iron plates was built up, to provide form for the 2-foot, permanent brick lining that followed. The three stages are shown in the model in about their proper relationship of progress. The work is shown passing beneath an old timber-crib bulkhead, used for stabilizing the shoreline.

The silt of the riverbed was about the consistency of putty and under good conditions formed a secure barrier between the excavation and the river above. It was easily excavated, and for removal was mixed with water and blown out through a pipe into the shaft by the higher pressure in the tunnel. About half was left in the bore for removal later. The basic scheme was workable, but in operation an extreme precision was required in regulating the air pressure in the work area. [5] It was soon found that there existed an 11-psi difference between the pressure of water on the top and the bottom of the working face, due to the 22-foot height of the unlined opening. Thus, it was impossible to maintain perfect pneumatic balance of the external pressure over the entire face. It was necessary to strike an average with the result that some water entered at the bottom of the face where the water pressure was greatest, and some air leaked out at the top where the water pressure was below the air pressure. Constant attention was essential: several men did nothing but watch the behavior of the leaks and adjusted the pressure as the ground density changed with advance. Air was supplied by several steam-driven compressors at the surface.

The air lock permitted passage back and forth of men and supplies between the atmosphere and the work area, without disturbing the pressure differential. This principle is demonstrated by an animated model set into the main model, to the left of the shaft (fig. 39). The variation of pressure within the lock chamber to match the atmosphere or the pressurized area, depending on the direction of passage, is clearly shown by simplified valves and gauges, and by the use of light in varying color density. In the Haskin tunnel, 5 to 10 minutes were taken to pass the miners through the lock so as to avoid too abrupt a physiological change.

Despite caution, a blowout occurred in July 1880 due to air leakage not at the face, but around the temporary entrance. One door of the air lock jammed and twenty men drowned, resulting in an inquiry which brought forth much of the distrust with which Haskin was regarded by the engineering profession. His ability and qualifications were subjected to the bitterest attack in and by the technical press. There is some indication that, although the project began with a staff of competent engineers, they were alienated by Haskin in the course of work and at least one withdrew. Haskin's remarks in his own defense indicate that some of the denunciation was undoubtedly justified. And yet, despite this reaction, the fundamental merit of the pneumatic tunneling method had been demonstrated by Haskin and was immediately recognized and freely acknowledged. It was apparent at the same time, however, that air by itself did not provide a sufficiently reliable support for large-area tunnel works in unstable ground, and this remains the only major subaqueous tunnel work driven with air alone.

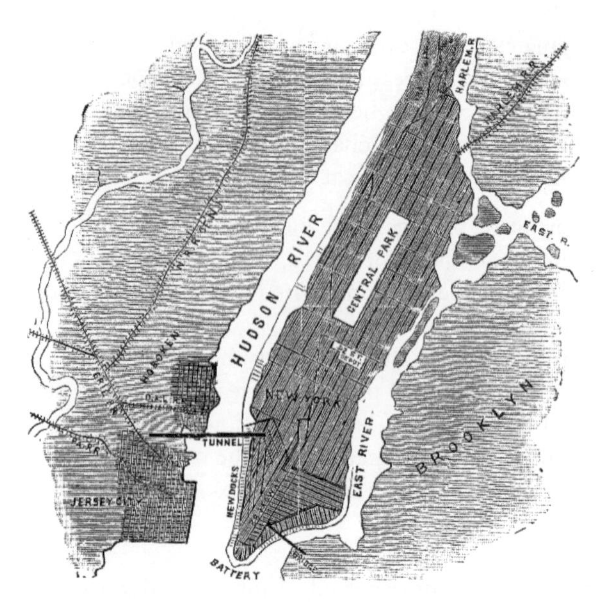

Figure 41.—LOCATION OF HUDSON RIVER TUNNEL. (*Leslie's Weekly*, 1879.)

After the accident, work continued under Haskin until 1882 when funds ran out. About 1600 feet of the north tube and 600 feet of the south tube had been completed. Greathead resumed operations with a shield for a British company in 1889, but exhaustion of funds again caused stoppage in 1891. The tunnel was finally completed in 1904, and is now in use as part of the Hudson and Manhattan rapid-transit system, never providing the sought-after rail link. A splendid document of the Haskin portion of the work is S. D. V. Burr's *Tunneling Under the Hudson River* published in 1885. It is based entirely upon firsthand material and contains drawings of most of the work, including the auxiliary apparatus. It is interesting to note that electric illumination (arc, not incandescent, lights) and telephones were used, unquestionably the first employment of either in tunnel work.

Figure 42.—ST. CLAIR TUNNEL. View of front of shield showing method of excavation in firm strata. Incandescent electric illumination was used. 1889-90. MHT model—1" scale. (Smithsonian photo 49260-D.)

THE ST. CLAIR TUNNEL

The final model of the soft-ground series reflects, as did the Hoosac Tunnel model for hard-rock tunneling, final emergence into the modern period. Although the St. Clair Tunnel was completed over 70 years ago, it typifies in its method of construction, the basic procedures of subaqueous work in the present day. The Thames Tunnel of Brunel, and Haskin's efforts beneath the Hudson, had clearly shown that by themselves, both the shield and pneumatic systems of driving through fluid ground were defective in practice for tunnels of large area. Note that the earliest successful works by each method had been of very small area, so that the influence of adverse conditions was greatly diminished.

The first man to perceive and seize upon the benefits to be gained by combining the two systems was, most fittingly, Greathead. Although he had projected the technique earlier, in driving the underground City and South London Railway in 1886, he brought together for the first time the three fundamental elements essential for the practical tunneling of soft, water-bearing ground: compressed-air support of the work during construction, the movable shield, and cast-iron, permanent lining. The marriage was a happy one indeed; the limitations of each system were almost perfectly overcome by the qualities of the others.

The conditions prevailing in 1882 at the Sarnia, Ontario, terminal of the Grand Trunk Railway, both operational and physical, were almost precisely the same as those which inspired the undertaking of the Hudson River Tunnel. The heavy traffic at this vital U.S.—Canada rail interchange was ferried inconveniently across the wide St. Clair River, and the bank and river conditions precluded construction of a bridge. A tunnel was projected by the railway in that year, the time when Haskin's tribulations were at their height. Perhaps because of this lack of precedent for a work of such size, nothing was done immediately. In 1884 the railway organized a tunnel company; in 1886 test borings were made in the riverbed and small exploratory drifts were started across from both banks by normal methods of mine timbering. The natural gas, quicksand, and water encountered soon stopped the work.

Figure 43.—REAR VIEW OF ST. CLAIR SHIELD showing the erector arm placing a cast-iron lining segment. The three motions of the arm—axial, radial, and rotational, were manually powered. (Smithsonian photo 49260-C.)

It was at this time that the railway's president visited Greathead's City and South London workings. The obvious answer to the St. Clair problem lay in the successful conduct of this subway. Joseph Hobson, chief engineer of the Grand Trunk and of the tunnel project, in designing a shield, is said to have searched for drawings of the shields used in the Broadway and Tower Subways of 1868-9, but unable to locate any, he relied to a limited extent on the small drawings of those in Drinker's volume. There is no explanation as to why he did not have drawings of the City and South London shield at that moment in use, unless one considers the rather unlikely possibility that Greathead maintained its design in secrecy.

Figure 44.—OPENING OF THE ST. CLAIR TUNNEL, 1891. (*Photo courtesy of Detroit Library, Burton Historical Collection.*)

The Hobson shield followed Greathead's as closely as any other, in having a diaphragm with closable doors, but a modification of Beach's sharpened horizontal shelves was also used. However, these functioned more as working platforms than supports for the earth. The machine was 21½ feet in diameter, an unprecedented size and almost twice that of Greathead's current one. It was driven by 24 hydraulic rams. Throughout the entire preliminary consideration of the project there was a marked sense of caution that amounted to what seems an almost total lack of confidence in success. Commencement of the work from vertical shafts was planned so that if the tunnel itself failed, no expenditure would have been made for approach work. In April 1888, the shafts were started near both riverbanks, but before reaching proper depth the almost fluid clay and silt flowed up faster than it could be excavated and this plan was abandoned. After this second inauspicious start, long open approach cuts were made and the work finally began. The portals were established in the cuts, several thousand feet back from each bank and there the tunneling itself began. The portions under the shore were driven without air. When the banks were reached, brick bulkheads containing air locks were built across the opening and the section beneath the river, about 3,710 feet long, driven under air pressure of 10 to 28 pounds above atmosphere. For most of the way, the clay was firm and there was little air leakage. It was found that horses could not survive in the compressed air, and so mules were used under the river.

In the firm clay, excavation was carried on several feet in front of the shield, as shown in the model (fig. 42). About twelve miners worked at the face. However, in certain strata the clay encountered was so fluid that the shield could be simply driven forward by the rams, causing the muck to flow in at the door openings without excavation. After each advance, the rams were retracted and a ring of iron lining segments built up, as in the Tower Subway. Here, for the first time, an "erector arm" was used for placing the segments, which weighed about half a ton. In all respects, the work advanced with wonderful facility and lack of operational difficulty. Considering the large area, no subaqueous tunnel had ever been driven with such speed. The average monthly progress for the American and Canadian headings totaled 455 feet, and at top efficiency 10 rings or a length of 15.3 feet could be set in a 24-hour day in each heading. The 6,000 feet of tunnel was driven in just a year; the two shields met vis-a-vis in August of 1890.

The transition was complete. The work had been closely followed by the technical journals and the reports of its successful accomplishment thus were brought to the attention of the entire civil engineering profession. As the first major subaqueous tunnel completed in America and the first in the world of a size able to accommodate full-scale rail traffic, the St. Clair Tunnel served to dispel the doubts surrounding such work, and established the pattern for a mode of tunneling which has since changed only in matters of detail.

Of the eight models, only this one was built under the positive guidance of original documents. In the possession of the Canadian National Railways are drawings not only of all elements of the shield and lining, but of much of the auxiliary apparatus used in construction. Such materials rarely survive, and do so in this case only because of the foresight of the railway which, to avoid paying a high profit margin to a private contractor as compensation for the risk and uncertainty involved, carried the contract itself and, therefore, preserved all original drawing records.

While the engineering of tunnels has been comprehensively treated in this paper from the historical standpoint, it is well to still reflect that the advances made in tunneling have not perceptibly removed the elements of uncertainty but have only provided more positive and effective means of countering their forces. Still to be faced are the surprises of hidden streams, geologic faults, shifts of strata, unstable materials, and areas of extreme pressure and temperature.

BIBLIOGRAPHY

AGRICOLA, GEORGIUS. *De re Metallica.* [English transl. H. C. and L. H. Hoover (*The Mining Magazine*, London, 1912).] Basel: Froben, 1556.

BEACH, ALFRED ELY. *The pneumatic dispatch.* New York: The American News Company, 1868.

BEAMISH, RICHARD. *A memoir of the life of Sir Marc Isambard Brunel.* London: Longmans, Green, Longmans and Roberts, 1862.

BURR, S. D. V. *Tunneling under the Hudson River.* New York: John Wiley and Sons, 1885.

COPPERTHWAITE, WILLIAM CHARLES. *Tunnel shields and the use of compressed air in subaqueous works.* New York: D. Van Nostrand Company, 1906.

DRINKER, HENRY STURGESS. *Tunneling, explosive compounds and rock drills.* New York: John Wiley and Sons, 1878.

LATROBE, BENJAMIN H. Report on the Hoosac Tunnel (Baltimore, October 1, 1862). Pp. 125-139, app. 2, in *Report of the commissioners upon the Troy and Greenfield Railroad and Hoosac Tunnel.* Boston, 1863.

LAW, HENRY. A memoir of the Thames Tunnel. *Weale's Quarterly Papers on Engineering* (London, 1845-46), vol. 3, pp. 1-25 and vol. 5, pp. 1-86.

The pneumatic tunnel under Broadway, N.Y. *Scientific American* (March 5, 1870), pp. 154-156.

Report of the commissioners upon the Troy and Greenfield Railroad and Hoosac Tunnel to his excellency the governor and the honorable the executive council of the state of Massachusetts, February 28, 1863. Boston, 1863.

STORROW, CHARLES S. Report on European tunnels (Boston, November 28, 1862). Pp. 5-122, app. 1, in *Report of the commissioners upon the Troy and Greenfield Railroad and Hoosac Tunnel....* Boston, 1863.

The St. Clair Tunnel. *Engineering News* (in series running October 4 to December 27, 1890).

FOOTNOTES:

[1] There are two important secondary techniques for opening subterranean and subaqueous ways, neither a method truly of tunneling. One of these, of ancient origin, used mainly in the construction of shallow subways and utility ways, is the "cut and cover" system, whereby an open trench is excavated and then roofed over. The result is, in effect, a tunnel. The concept of the other method was propounded in the early 19th century but only used practically in recent years. This is the "trench" method, a sort of subaqueous equivalent of cut and cover. A trench is dredged in the bed of a body of water, into which prefabricated sections of large diameter tube are lowered, in a continuous line. The joints are then sealed by divers, the trench is backfilled over the tube, the ends are brought up to dryland portals, the water is pumped out, and a subterranean passage results. The Chesapeake Bay Bridge Tunnel (1960-1964) is a recent major work of this character.

[2] In 1952 a successful machine was developed on this plan, with hardened rollers on a revolving cutting head for disintegrating the rock. The idea is basically sound, possessing advantages in certain situations over conventional drilling and blasting systems.

[3] In 1807 the noted Cornish engineer Trevithick commenced a small timbered drift beneath the Thames, 5 feet by 3 feet, as an exploratory passage for a larger vehicular tunnel. Due to the small frontal area, he was able to successfully probe about 1000 feet, but the river then broke in and halted the work. Mine tunnels had also reached beneath the Irish Sea and various rivers in the coal regions of Newcastle, but these were so far below the surface as to be in perfectly solid ground and can hardly be considered subaqueous workings.

[4] Unlike the Brunel tunnel, this was driven from both ends simultaneously, the total overall progress thus being 3 feet per shift rather than 18 inches. A top speed of 9 feet per day could be advanced by each shield under ideal conditions.

[5] Ideally, the pressure of air within the work area of a pneumatically driven tunnel should just balance the hydrostatic head of the water without, which is a function of its total height above the opening. If the air pressure is not high enough, water will, of course, enter, and if very low, there is danger of complete collapse of the unsupported ground areas. If too high, the air pressure will overcome that due to the water and the air will force its way out through the ground, through increasingly larger openings, until it all rushes out suddenly in a "blowout." The pressurized atmosphere gone, the water then is able to pour in through the same opening, flooding the workings.

Milton Keynes UK
Ingram Content Group UK Ltd.
UKHW030846141124
451205UK00005B/466